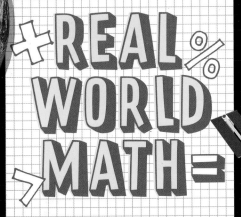

REAL WORLD MATH

SPACE EXPLORATION

by Jennifer Szymanski

9 ... 8 ...

10 ...

Children's Press®
An imprint of Scholastic Inc.

Library of Congress Cataloging-in-Publication Data
Names: Szymanski, Jennifer, author.
Title: Space Exploration / Jennifer Szymanski.
Description: First edition. | New York : Children's Press, an
imprint of Scholastic Inc. 2021. | Series: Real world math |
Includes index. | Audience: Ages 5–7. | Audience: Grades K–1. |
Summary: "This book introduces young readers to math
concepts around space exploration"— Provided by publisher.
Identifiers: LCCN 2021000033 (print) | LCCN 2021000034 (ebook) |
ISBN 9781338762396 (library binding) | ISBN 9781338762402
(paperback) | ISBN 9781338762419 (ebook)
Subjects: LCSH: Mathematics—Juvenile literature. | Space
sciences—Mathematics—Juvenile literature. |
Moon—Exploration—Juvenile literature. | Mars
(Planet)—Exploration—Juvenile literature. | Outer
space—Exploration—Juvenile literature.
Classification: LCC QA135.6 .S98 2021 (print) |
LCC QA135.6 (ebook) | DDC 520.1/51—dc23
LC record available at https://lccn.loc.gov/2021000033
LC ebook record available at https://lccn.loc.gov/2021000034

10 9 8 7 6 5 4   23 24 25 26

Printed in the
U.S.A. 40
First edition, 2022

Series produced by WonderLab Group, LLC
Book design by Moduza Design
Photo editing by Annette Kiesow
Educational consulting by Leigh Hamilton
Copyediting by Vivian Suchman
Proofreading by Molly Reid
Indexing by Connie Binder

Photos ©: back cover center: JPL-Caltech/NASA; 1 top left,
top right: NASA; 2 top right: NASA/Goddard/SDO/Flickr; 2
bottom left: Provectorstock/Dreamstime; 2 bottom right
inset: NASA/ESA/STScI; 3 bottom left: JPL-Caltech/NASA;
4–5 top: PIA2NASA/JPL-Caltech; 5 center: Courtesy of Eva L.
Scheller; 6–7: Hannu Viitanen/Dreamstime; 8 inset:
Provectorstock/Dreamstime; 8: Scott Andrews/NASA;
9: NASA/JPL-Caltech; 11 John Kaufmann/NASA; 12: NASA/
Goddard/Arizona State University; 13 left: NASA/JSC;
13 center left: NASA/JPL; 13 center right, right: NASA; 14–15:
NASA; 15 inset: NASA/Roscosmos; 16: NASA; 18–19: NASA/JPL/
University of Arizona/USGS/Kevin M. Gill/Flickr; 18 inset
left, 18 inset right, 19 inset: NASA/JPL/MSSS; 20 top right,
bottom right: NASA/JPL-Caltech/MSSS; 20 center: NASA/
JPL-Caltech/MSSS/Kevin M. Gill/Flickr; 20 bottom left:
JPL-Caltech/NASA; 21 top left: NASA/JPL/University of
Arizona/USGS/Kevin M. Gill/Flickr; 21 top right, center left,
center right, bottom left, bottom right: NASA/JPL-Caltech/
MSSS; 22–23: Zhasminaivanova/Dreamstime; 22 inset: NASA/
JPL-Caltech; 24: NASA; 26: GSFC/NASA; 27 top: NASA/JSC; 27
bottom: NASA/JPL-Caltech; 28 all: Courtesy of Eva L.
Scheller; 29 top: NASA/JPL-Caltech/MSSS; 29 bottom: NASA/JPL.

All other photos © Shutterstock.

CONTENTS

LET'S GO!

3…2…1… it's time to blast off!

Let's hop aboard a rocket. Make sure to grab a space suit and your math skills!

Astronomers are scientists who study space. Math helps astronomers learn new things about **planets**, stars, and other objects in the sky. It also helps them build rockets. They can **measure** the distance between objects in space. They can study the size of **craters** on the moon. And astronomers can use **addition** to make sure they have the right amount of equipment on their rocket.

Today we are taking a trip to explore space using math. Before we launch, let's look at the planets in our **solar system**. Are you ready?

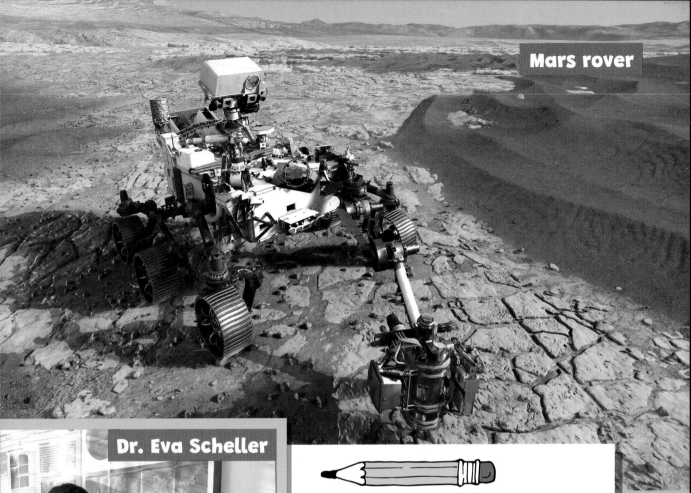

Mars rover

Dr. Eva Scheller

MEET DR. EVA

We are joining Dr. Eva Scheller on our journey. She is an astronomer who studies the planet Mars. She looks for answers to questions like "What did Mars look like a long time ago?" Dr. Eva also studies the **geology** of Mars. This means she studies its rocks and lakes. Her research can help us find out what other planets are like.

1 Mercury
2 Venus
3 Earth
4 Mars
5 Jupiter

At night, the sky is filled with twinkling lights. Most of the lights we see from Earth are stars. Some are planets. For a long time, scientists have wondered what it is like on other planets.

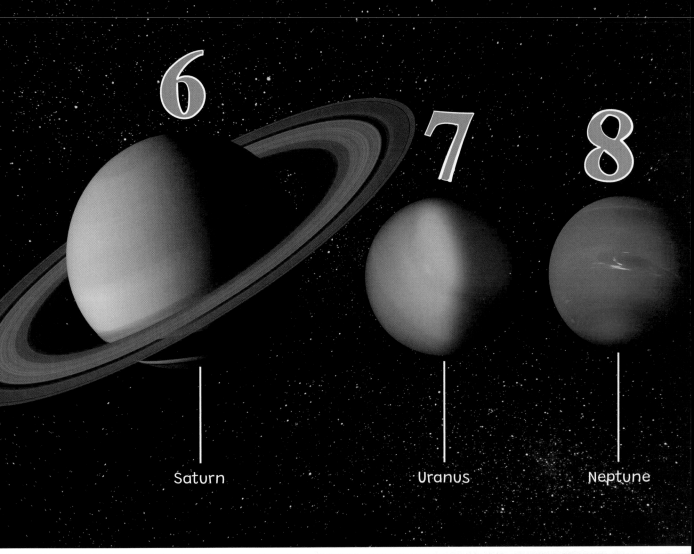

6

7

8

Saturn

Uranus

Neptune

Scientists often count the planets beginning with Mercury. Mercury is the planet closest to the sun. Neptune is the farthest. Now we know more about our solar system. So let's hop aboard our rocket!

It is almost time for launch.

When the rocket is ready to go, there is a countdown. A clock ticks down the seconds until blastoff. When ten seconds are left, someone in mission control will start counting backward from ten to one.

10 SEC

YOU CAN DO IT! ①

Help the rocket blast off. Count backward starting from ten.

10 9 8 7 6 5 4 3 2 1

Blast off!

We're on our way! Let's get an up-close look at a familiar round object. You can see it from Earth almost every night.

Earth's moon is coming into view!

Back on Earth, you can sometimes see the shape of a face in the moon. But up close, we see that there is no face at all. The moon's surface is dotted with holes called craters. Craters form when an object from space smashes into the moon.

EXPLORE WITH DR. EVA

When an object hits the moon, rocks fly from the crater that is created. Scientists like Dr. Eva carefully examine the size and shape of these rocks. They often give scientists clues about what the object was made of or where it came from.

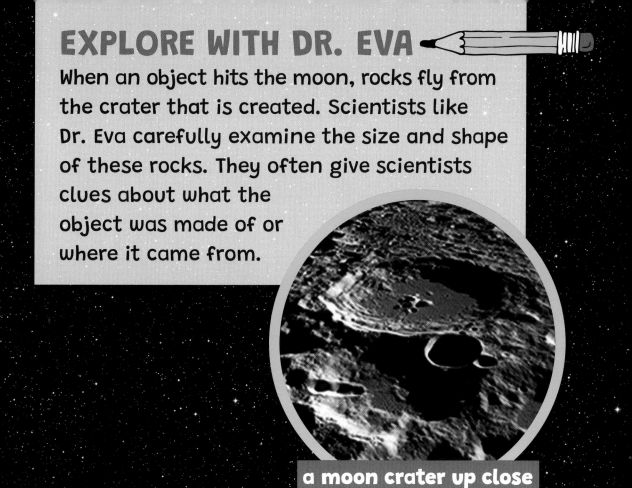

a moon crater up close

Craters can be big or small. Scientists can learn a lot about the object that hit the moon from the crater it makes. They look at the size of the hole. And they look at the size of the rocks around it. Big craters are usually made by big objects, like huge asteroids. Smaller craters are sometimes made by small pieces of asteroids, called meteoroids.

The moon has thousands of craters. Astronomers named this one Tycho (tie-koh) crater. It formed long ago when a large object hit the moon. The object blasted rocks far from Tycho.

YOU CAN DO IT! ②

Let's look at these rocks that came from the moon. Putting the rocks in order by size can help us learn more about them. Can you put them in order from largest to smallest?

A   B   C   D

0 1 2 3 4 5 6 7

 Good job! Let's check in with some space explorers.

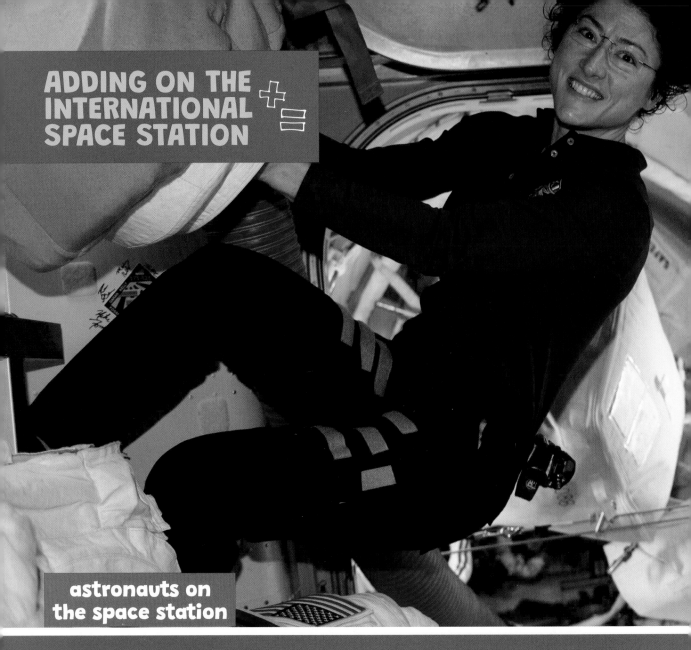

ADDING ON THE INTERNATIONAL SPACE STATION ＋ ＝

astronauts on the space station

Lights flash, machines beep, and fans whir. An astronaut's day is never boring. There is so much to do when you live in space. Some astronauts live in space for months aboard the International Space Station (ISS).

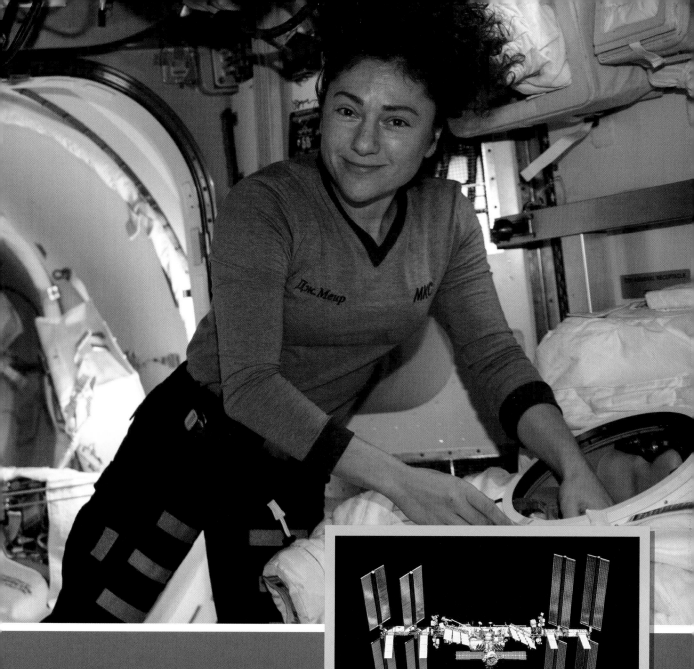

The ISS is a huge research station that **orbits** Earth.

International Space Station

It is almost as big as a football field. Life inside the ISS is very busy. Astronauts exercise every day. They talk to researchers on Earth. And they work on science experiments.

15

Sometimes astronauts have to repair the ISS. Outside the station, astronauts wear special space suits. These keep astronauts safe while they do their job. Usually only one astronaut works outside of the ISS at a time. But to make this repair, two more astronauts are needed. One astronaut plus two astronauts equals three astronauts. We use addition to find out how many astronauts are working in all.

1 + 2 = 3

YOU CAN DO IT!

It's lunchtime. An astronaut's food comes in foil pouches. Each astronaut adds water to their food and heats it. Then it's time to eat.

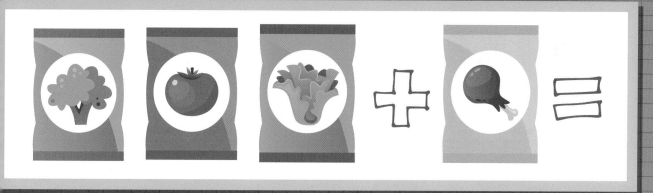

Healthy meals are important, even in space. Today's meal has three pouches of vegetables and one pouch of chicken. Now add them up.

How many pouches of food are in today's meal?

 Great work adding! Look—a planet is coming into view.

CATEGORIES ON MARS ○△

ice on Mars

Mars rover

Mars is sometimes called the "Red Planet." This is because it is covered in red, dusty soil. Mars has volcanoes, mountains, and ice. Scientists study Mars to find out if people could visit Mars or live there someday. For now, only robots called **rovers** have landed on the planet's surface.

EXPLORE WITH DR. EVA

Mars rovers carry more than just cameras. Some rovers carry drills. Drills help rovers get samples of rocks and soil. Computers on the rovers can test the soil samples to see what it is made of. Dr. Eva studies samples like these to learn more about the planet.

rocks on Mars

Rovers can send back pictures for scientists to study. Scientists can sort these pictures by putting them into different groups called **categories**. Categories help scientists learn more about different parts of the planet.

Mars rovers can send hundreds of pictures back to Earth. Some pictures are of rocks. Others are of the sky. Sometimes rovers take selfies. Selfies are pictures of themselves! Each type of picture is in a different category.

rover selfie

sky

rocks

YOU CAN DO IT!

This rover has sent a lot of photos of Mars! Sort the photos of rocks, sky, and selfies into different groups. How many categories are there? How many photos are there in each category? Are there other ways to put these photos into categories?

Well done! Are you feeling adventurous? Let's explore beyond our solar system . . .

Milky Way galaxy

We can see many bright stars from Earth. Most of them make up our **galaxy**, the Milky Way. We share the universe with many stars. In fact, our sun is a star. To us on Earth, the stars look close together. But out in space, they are very far away from one another.

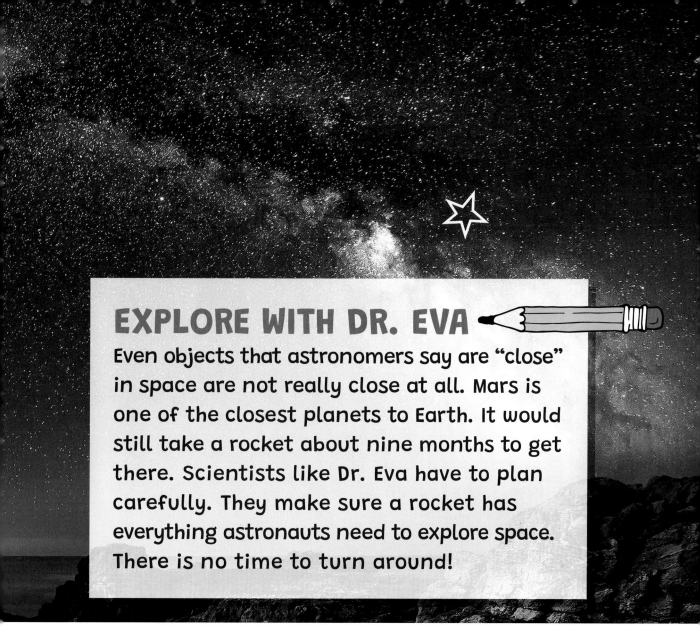

EXPLORE WITH DR. EVA

Even objects that astronomers say are "close" in space are not really close at all. Mars is one of the closest planets to Earth. It would still take a rocket about nine months to get there. Scientists like Dr. Eva have to plan carefully. They make sure a rocket has everything astronauts need to explore space. There is no time to turn around!

Stars are so far apart that scientists measure distance in a special way. Measuring the distance between stars can help scientists learn more about them. It can tell them how long it might take to travel there. It can even tell them how old the star might be.

The closest star to Earth outside of our solar system is called Proxima Centauri. The distance between this star and Earth is too large to measure like we would between other objects. So we will use a special extra-large **unit**. This will help us to understand the distance between Earth and Proxima Centauri.

There are **seven units** between Earth and Proxima Centauri.

YOU CAN DO IT! 5

Let's measure the distance between stars. How many units apart are these stars? Which set of stars are the farthest apart from each other?

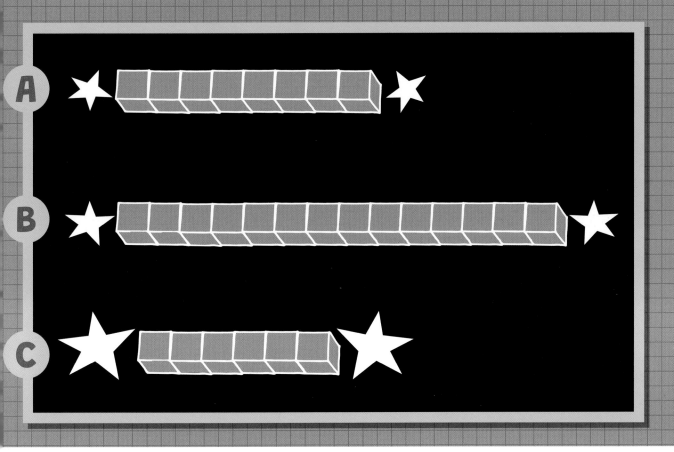

Hooray! We traveled far to visit the stars.
Time to head back to Earth.

WAY TO GO!

We did it! We explored parts of our solar system. We circled the moon. We met other astronauts. We watched rovers on the surface of Mars. And we traveled to faraway stars.

Earth's moon

As we explored space, we also learned how astronomers use math. We used counting, sizes, addition, categories, and measurement. These math skills help scientists learn more about our universe. Math is so important every day in so many ways. You might be surprised how often you use it in our world and beyond!

YOU CAN DO IT!

6

How many places did we explore on our trip? How many planets did we visit? If you could travel anywhere in space, where would you go first?

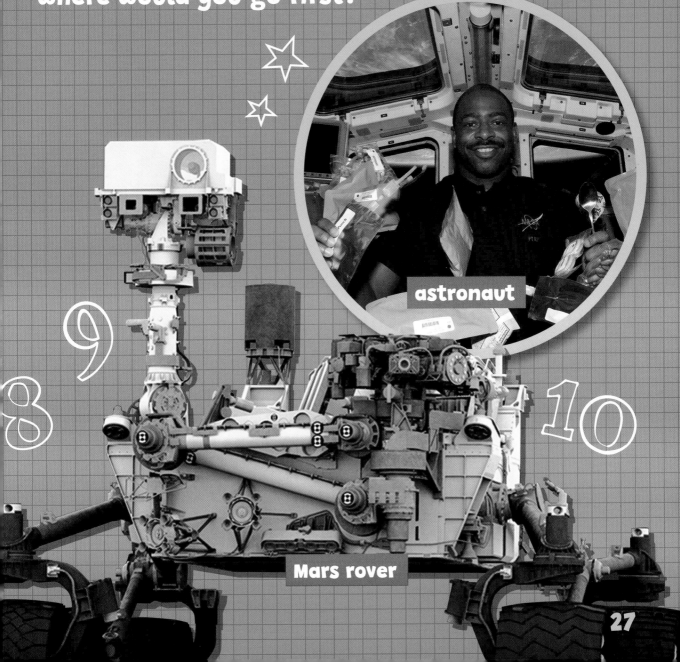

astronaut

Mars rover

DR. EVA EXPLORES

Dr. Eva Scheller is an astronomer.
She is also a geologist. She works at the California Institute of Technology. Some geologists study rocks and soil on Earth. But Dr. Eva studies rocks and soil on the surface of Mars!

People have never landed on Mars. But thanks to tools like computers, telescopes, and rovers, we know a lot about it. These tools take pictures and gather information about Mars. Then they send that information back to Earth.

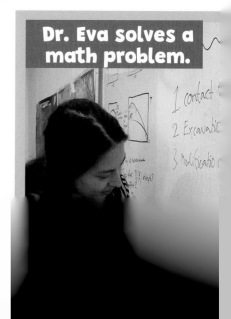

Dr. Eva solves a math problem.

Dr. Eva needs to use math to make sense of all that information. Thanks to math, Dr. Eva can figure out the size of a volcano on Mars. She can measure the distance between rock formations. She can find the shapes of old rivers or lakes. Dr. Eva can even count the number of days until the next rocket blasts off to Mars.

rocks from Mars

What Dr. Eva learns about space helps us learn more about our own planet. Thanks to math, Dr. Eva and other scientists are learning more every day. What will they discover next?

a satellite that takes photos of Mars

GLOSSARY

addition (uh-DISH-uhn): the combining of two or more numbers to come up with a sum

asteroid (AS-tuh-roid): a small, rocky object that travels around the sun

astronomer (uh-STRAH-nuh-mer): a scientist who studies stars, planets, and space

category (KAT-uh-gor-ee): a group of people or things that have certain characteristics in common

crater (KRAY-tur): a large hole in the ground caused by something falling or exploding

galaxy (GAL-uhk-see): a very large group of stars and planets

geology (jee-AH-luh-jee): the study of a planet's physical structure, especially its layers of soil and rock

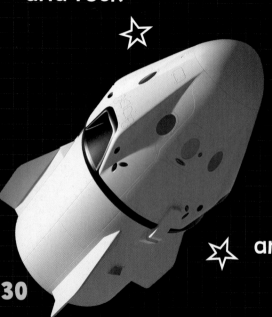

measure (MEZH-ur): to find out the size or weight of something

meteoroid (MEE-tee-uh-royd): a piece of rock in space formed after breaking off from an asteroid or other rocky object

YOU CAN DO IT! ANSWER KEY

① PAGE 9
10, 9, 8, 7, 6, 5, 4, 3, 2, 1, blast off!

② PAGE 13
D, A, C, B

③ PAGE 17
4 pouches

④ PAGE 21
There are three categories. There are two different frame colors: orange and yellow. This is also a category. There are two rover selfies, three rock photos, and one sky photo.

⑤ PAGE 25
A: 8 units; B: 14 units; C: 6 units; Set B has stars that are the farthest away from each other.

⑥ PAGE 27
We explored five places: the solar system, the moon, the ISS, Mars, and stars in the Milky Way. We visited one planet on our journey: Mars.

orbit (OR-bit): to travel in a circular path around something, especially a planet or the sun

planet (PLAN-it): one of the eight heavenly bodies circling the sun

rover (ROH-ver): a robot that explores the surface of a planet, like Mars

solar system (SOH-lur SIS-tuhm): the sun and all the things in its orbit, including planets, moons, asteroids, comets, and meteors

unit (YOO-nit): an amount used as a standard of measurement

INDEX

Page numbers in **bold** indicate illustrations.

ABOUT THE AUTHOR

Jennifer Szymanski is an author and freelance science education writer. She specializes in writing materials that support both teachers and students in meeting national and state science standards but considers her "real" job to be helping students connect science to everyday life. She lives near Pittsburgh, Pennsylvania, with her family and two cats.